THE GEARHEAD'S GUIDE TO CHOPPERS

BY LISA J. AMSTUTZ

CAPSTONE PRESS
a capstone imprint

Published by Spark, an imprint of Capstone
1710 Roe Crest Drive, North Mankato, Minnesota 56003
capstonepub.com

Copyright © 2023 by Capstone. All rights reserved. No part of this publication may be reproduced in whole or in part, or stored in a retrieval system, or transmitted in any form or by any means, electronic, mechanical, photocopying, recording, or otherwise, without written permission of the publisher.

Library of Congress Cataloging-in-Publication Data is available on the Library of Congress website
ISBN: 9781666356540 (hardcover)
ISBN: 9781666356557 (ebook PDF)

Summary: Choppers let riders take to the road in style. These motorcycles are the coolest things on two wheels. But there are lots of ways to make these vehicles faster and louder. Readers find out how in this high-interest book.

Editorial Credits
Editor: Erika L. Shores; Designer: Heidi Thompson; Media Researchers: Jo Miller and Pam Mitsakos; Production Specialist: Tori Abraham

Image Credits
Alamy: Georgiy Datsenko, 7, mauritius images GmbH, 9, Peter Brogden, 13, russ witherington, 11, Scott Hortop Travel, 5, Starsphinx, 28, Tim Gainey, 19, ZarkePix, 16; Shutterstock: Artoholics, 20, berdimm, 25, Brian McEntire, 14, Don Pablo, 15, i3alda, throughout, design element, Godlikeart, 8, James Steidl, 21, Ljupco Smokovski, 24, 29, n_defender, 23, vfhnb12, 22, Zoltan. Benyei, 27; Superstock: Mauritius, Cover

All internet sites appearing in back matter were available and accurate when this book was sent to press.

Capstone thanks Kevin Dick, technology education instructor in Mankato, MN, for his assistance in reviewing this book.

Table of Contents

READY TO ROLL 4

SPEED ZONE 6

MOVING PARTS 12

THE COOL FACTOR 18

    GLOSSARY 30

    READ MORE 31

    INTERNET SITES 31

    INDEX 32

    ABOUT THE AUTHOR 32

Ready to Roll

Rrrrrrrrrrr. You hear a low rumble. Here comes a chopper!

A chopper is a type of motorcycle. It has a long front end. Each chopper is one of a kind. Some people build the entire bike themselves. Other people start with a **stock** bike. They add **custom** parts.

> ## FACT
> Early choppers were called bobbers. Fenders were cut off, or "bobbed." Later, when major frame modifications were made, these "chopped" bikes were called choppers.

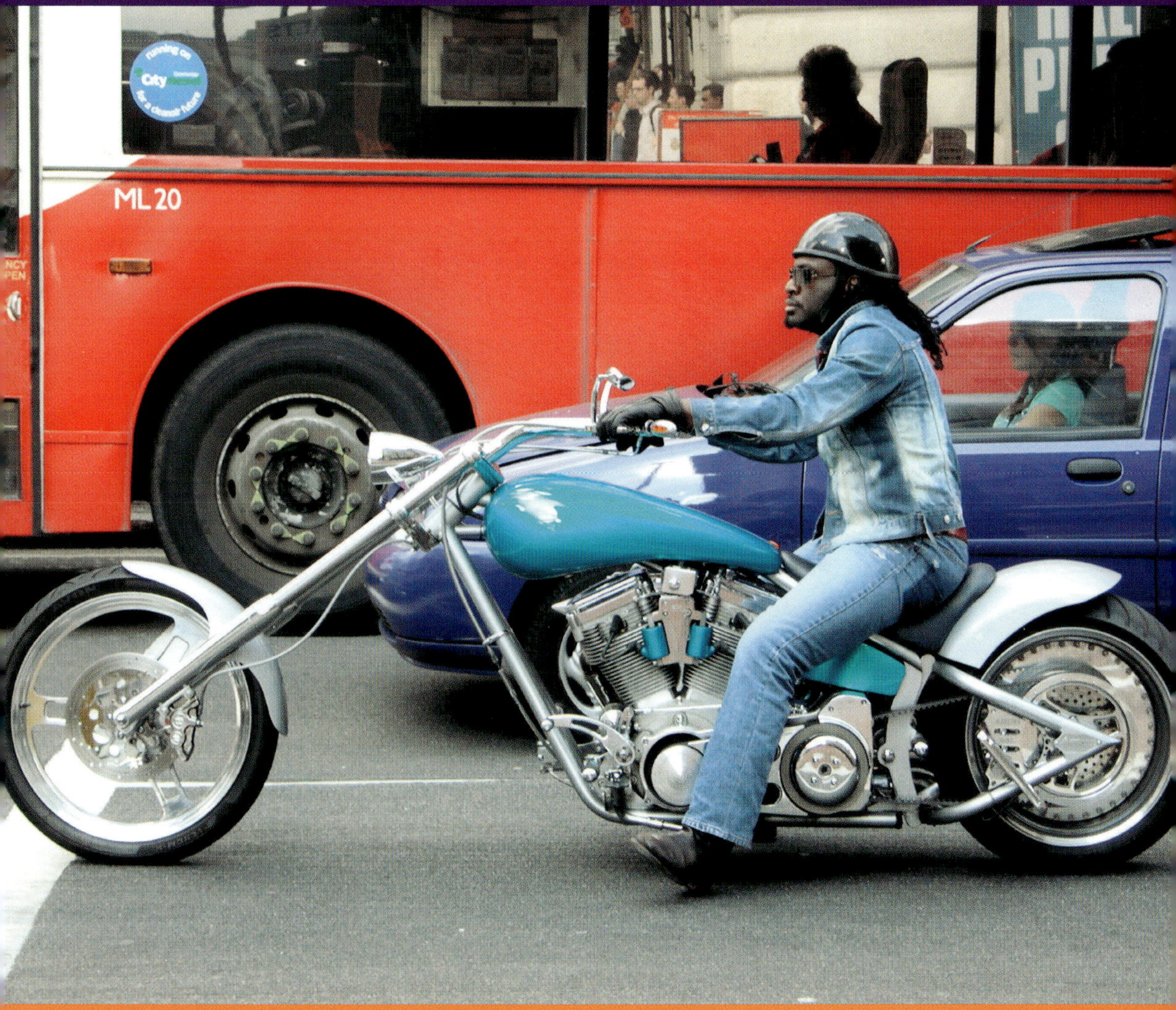
ML 20

Speed Zone

Choppers zoom down the road. Their top speed depends on engine size. One way to make them faster is swapping heavy parts for lighter ones. **Sprockets** are one place to start. **Aluminum** sprockets are lighter than steel.

FACT
Safety first! Riders must follow speed laws while on a chopper. A helmet and other safety gear are needed too.

sprocket

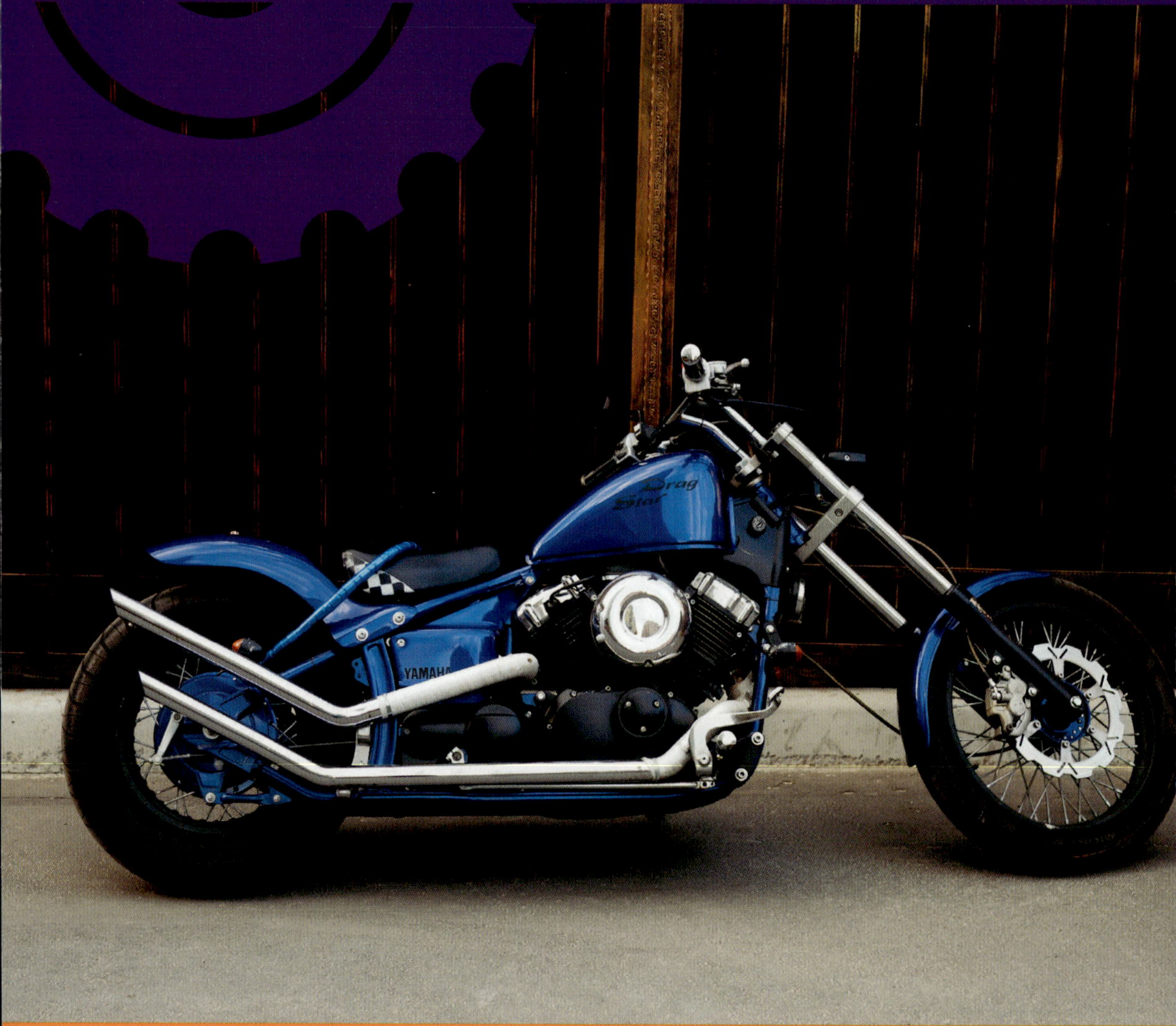
Drag
Star
YAMAHA

The **exhaust** system removes harmful gases. But stock systems can be heavy. A lighter one can free up more power. It can make the engine sound louder or deeper too. It makes that classic chopper sound.

Are the tires heavy or worn? Lighter ones will cut down on **friction**. They will boost the chopper's speed too. Lighter tires also do better on rough roads.

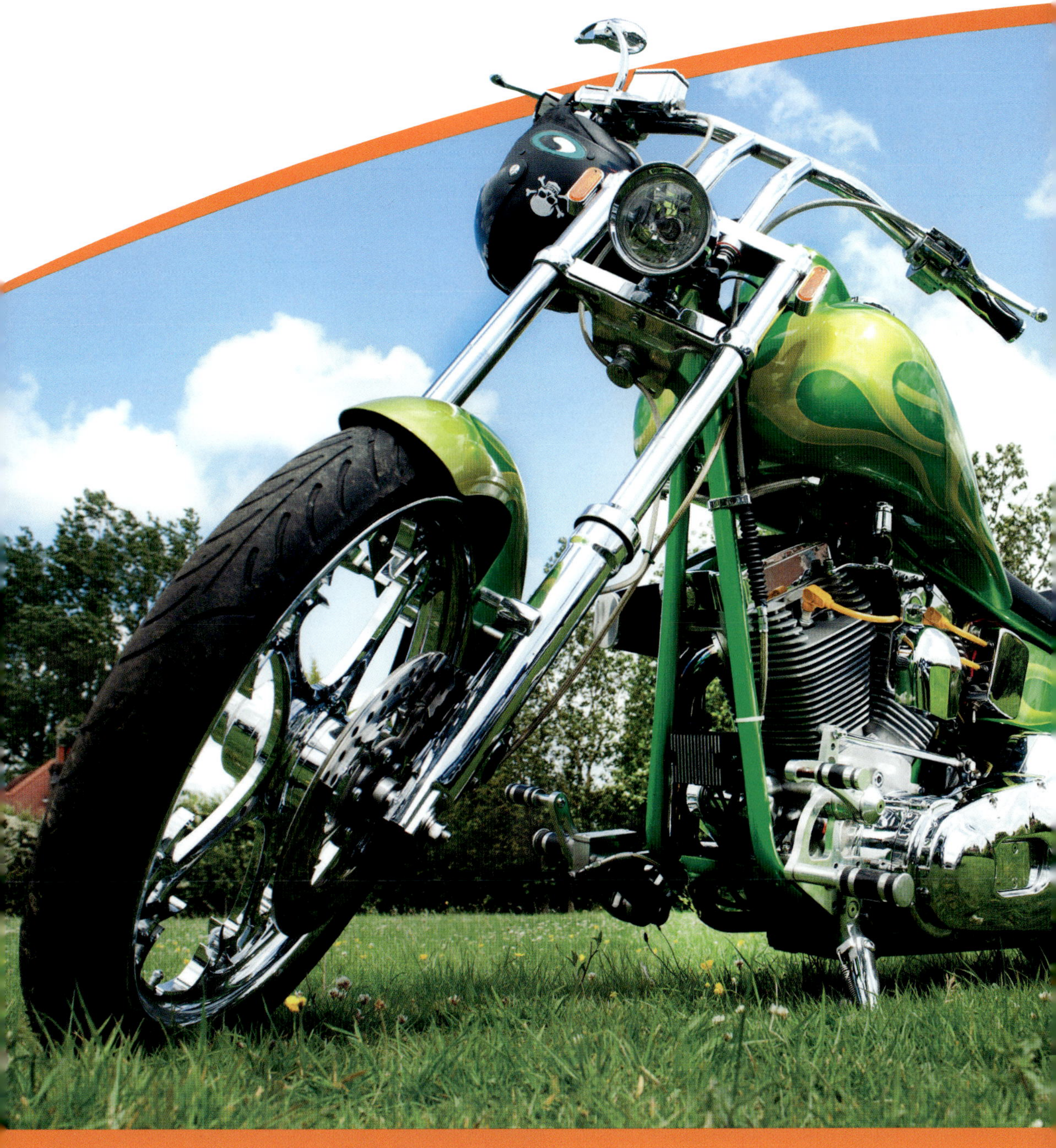

Moving Parts

Choppers look long and lean. They have long forks. The front tire is larger than the back tire. It has a small fender or none at all. Many have a tall sissy bar. This metal bar can be a backrest.

sissy bar

fork

Chopper parts are on display. There is no cover on them like there are on other motorcycles. Many chopper engines have a "V" shape. They are called V-twins.

Owners can rebuild an old engine. It will have that classic look.

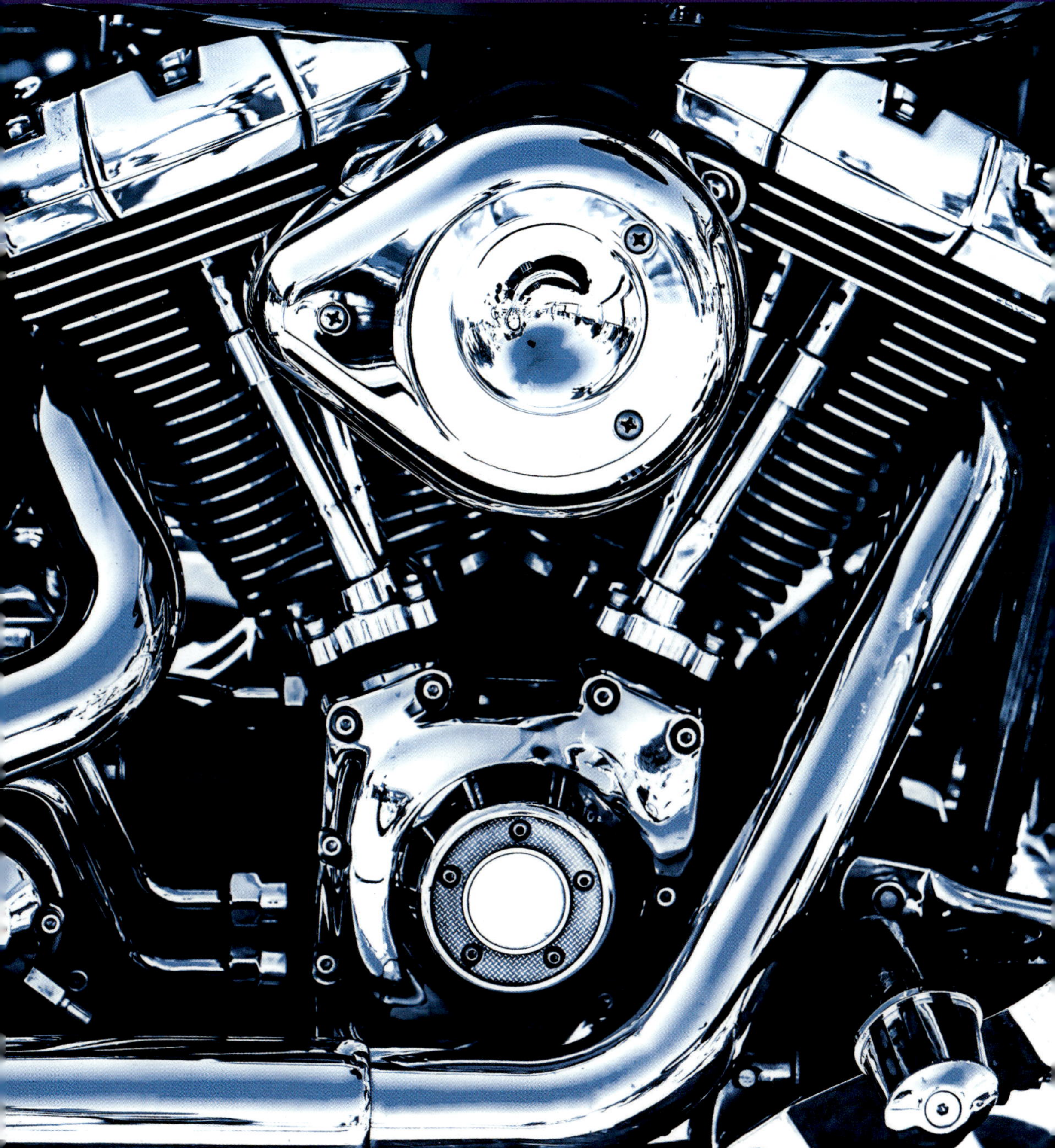

On sports cycles, riders sit forward. The foot pegs, brakes, and shift controls are below their body. But choppers are cruisers. The rider sits back. It helps to move those parts forward.

foot peg

FACT

A large motorcycle rally is held every year in Sturgis, South Dakota. Hundreds of thousands of bikes roll into town. They fill the streets.

The Cool Factor

Each chopper is different. That is what makes them cool. There are many tricks to make them even cooler!

Is the chrome scratched or chipped? A new coat of paint helps it look new again. Some owners get the chrome **engraved**. They add words or an image.

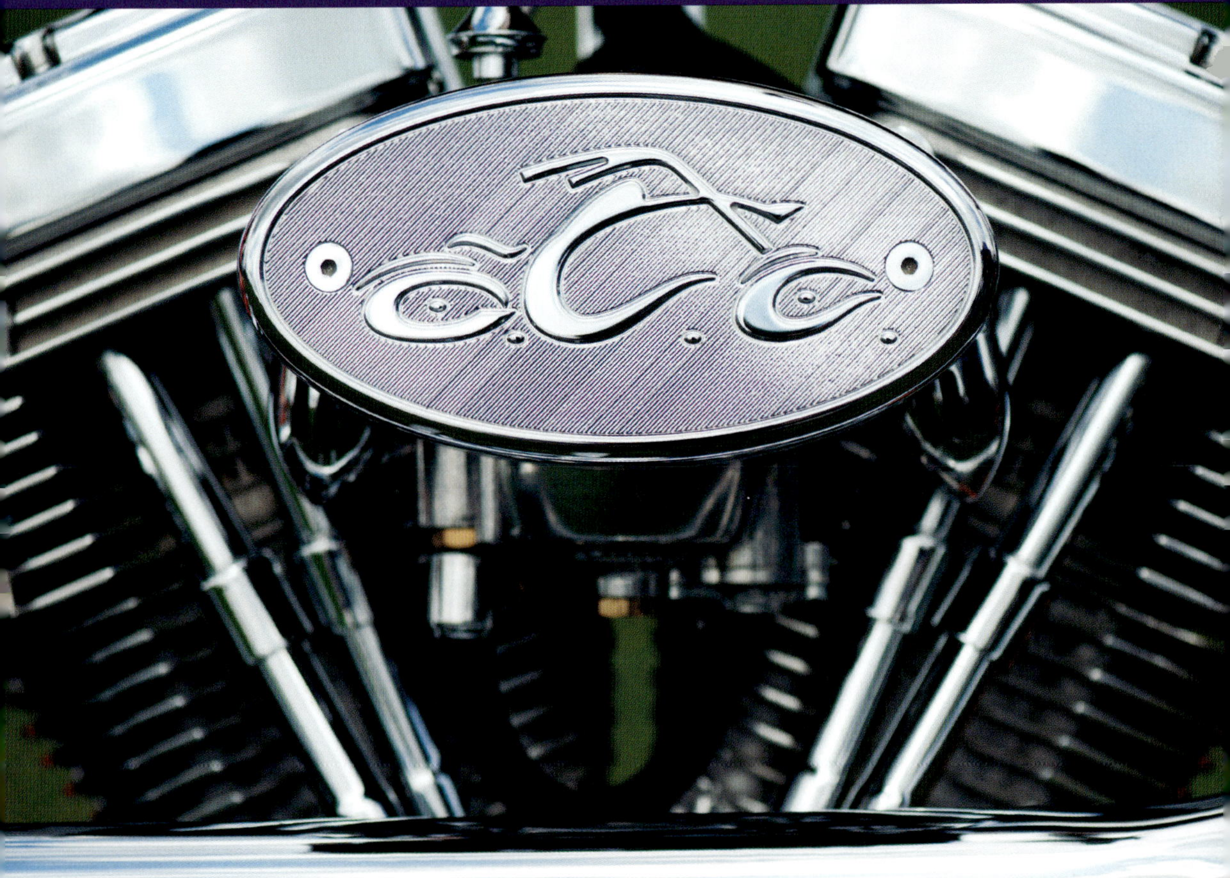

How can a chopper really stand out? Painting the frame and rims a bright color will do the trick! Or stick on **decals**. Racing stripes, stars, or flames are popular choices.

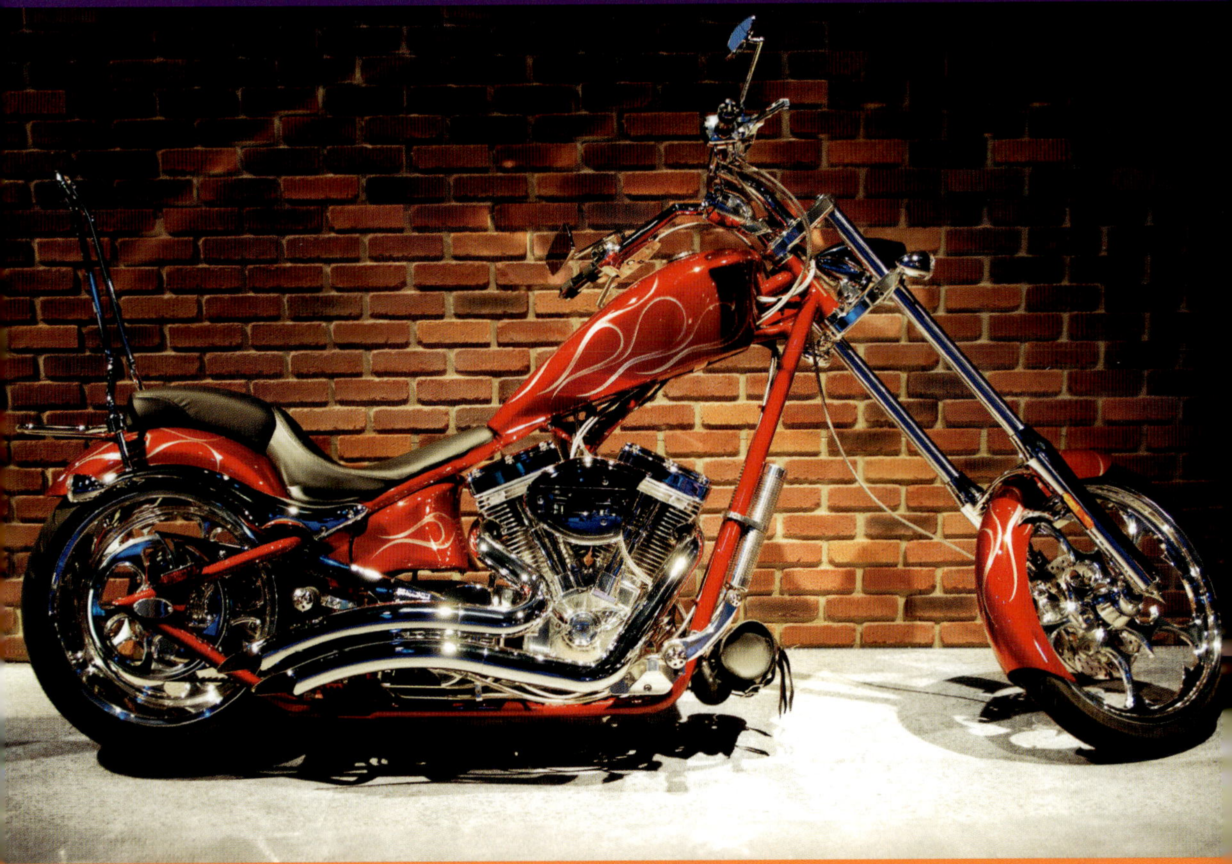

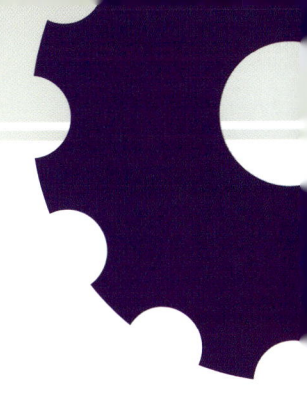

Owners might want to listen to some tunes. A radio and speakers can be added. Some riders add speakers or a headset to their helmets. But owners need to check state laws. These are not legal everywhere.

23

Grips are easy parts to upgrade. Owners might add heated ones. These grips keep hands warm in cold weather. A comfy pair is good for long rides.

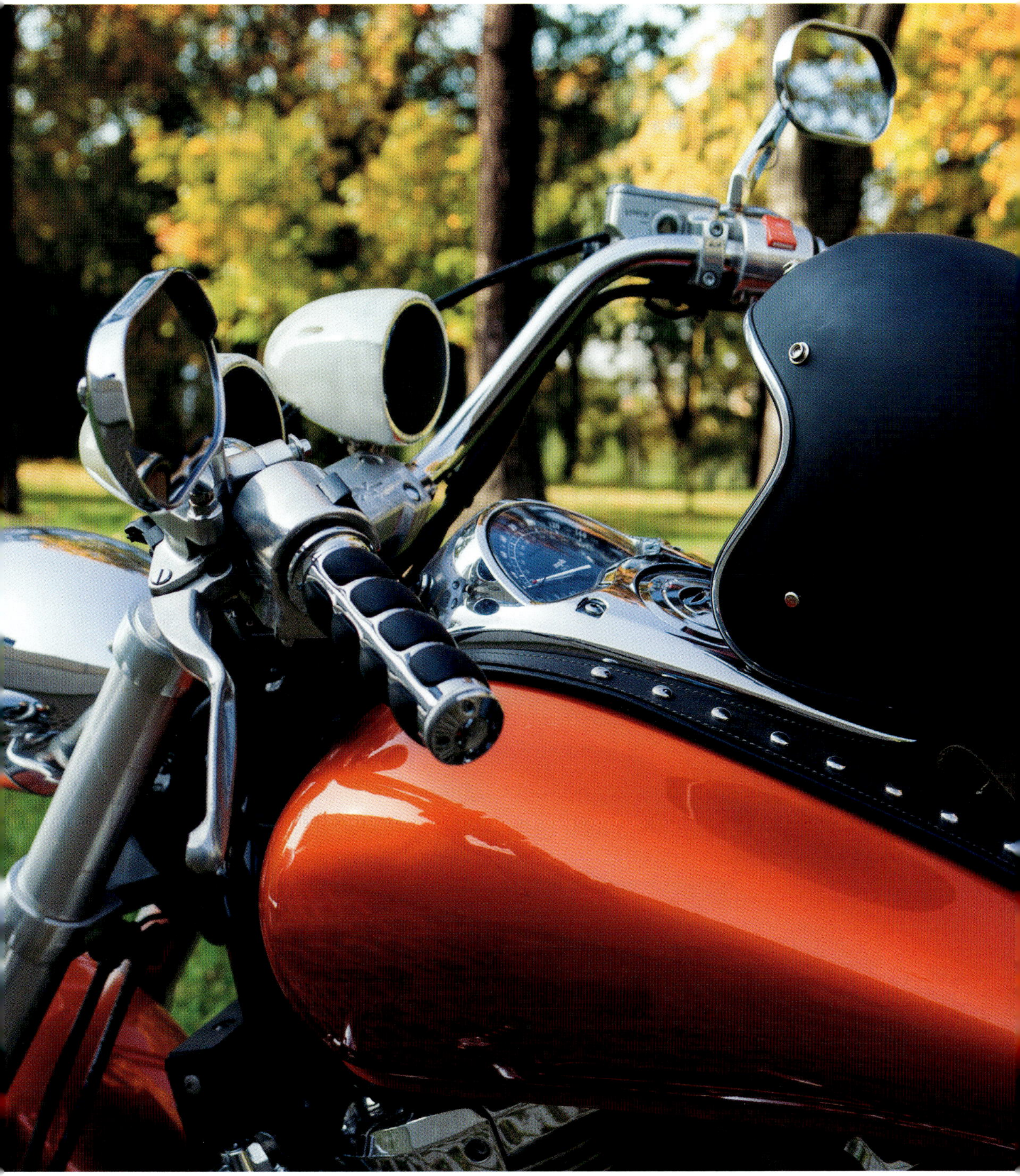

Light up the night with some strip or wheel lights. They make a chopper easier for other drivers to see. They look great too. Pencil beam driving lights make it easier for the rider to see in the dark. Fog lights can help in bad weather.

FACT
The 1969 movie *Easy Rider* showed actors on choppers. It made the bikes more popular than ever.

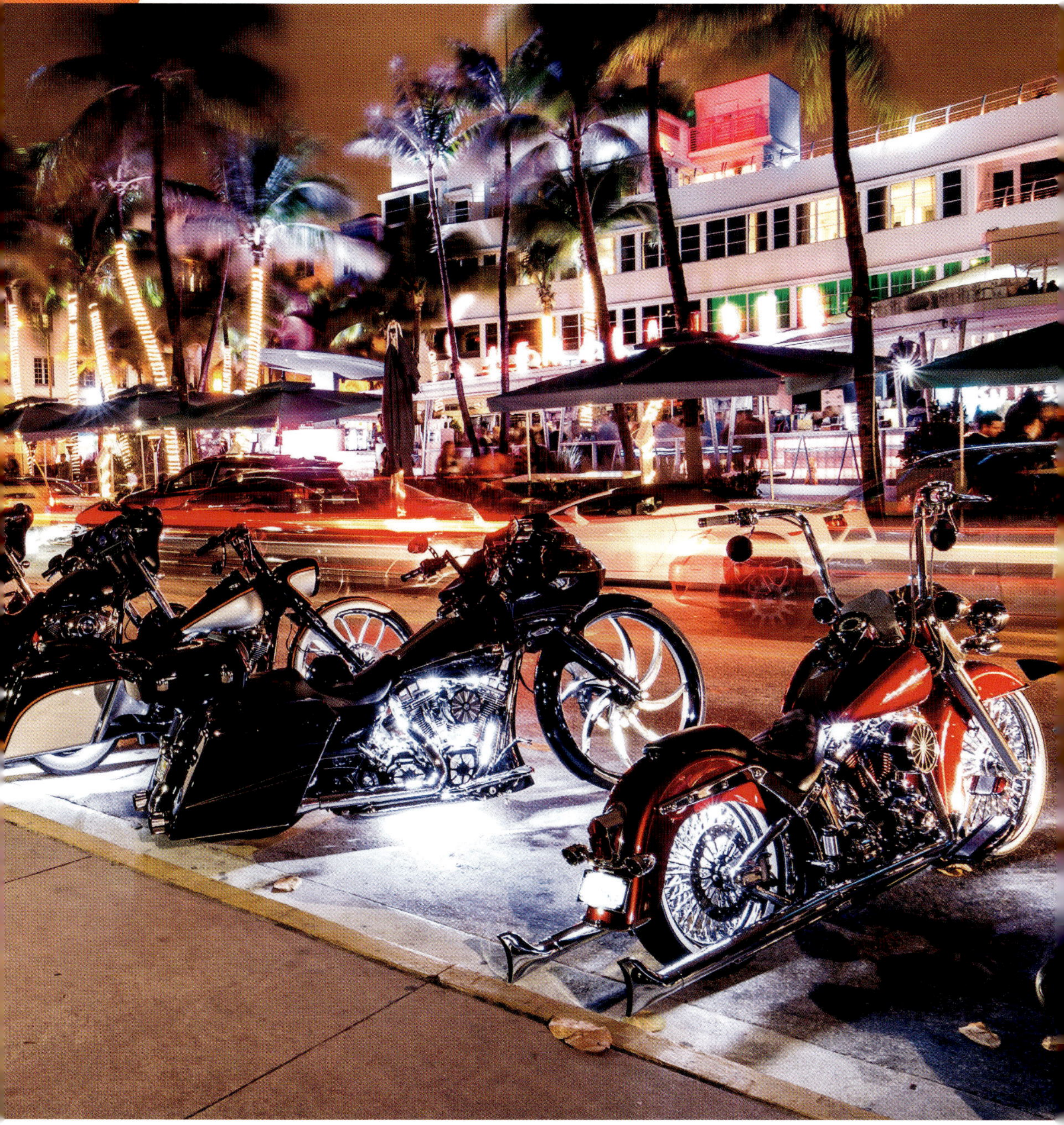

Want to cruise in comfort? Adding a backrest lets a rider lean back. A new seat cover will soften the bumps. Some choppers have a king and queen seat. It is built for two.

queen seat

king seat

Choppers are super cool and ready to roll. Sit back and enjoy the ride!

Glossary

aluminum (uh-LOO-muh-nuhm)—a lightweight metal

custom (KUHS-tuhm)—made to order

decal (DEE-kal)—a design printed on a sticker

engraved (in-GRAYVD)—cut into metal or another material

exhaust (eg-ZAWST)—the pipes that carry gases away from an engine

friction (FRIK-shuhn)—the force between two objects rubbing together

sprocket (SPROK-it)—a wheel with a rim of toothlike points that fit into the holes of a chain

stock (STOK)—the parts of a vehicle installed by the factory

Read More

Given-Wilson, Rachel, and Tamra B. Orr. *Your Future as an Auto Mechanic*. New York: The Rosen Publishing Group, Inc., 2020.

Motorbikes: 100 Extreme Machines. Tulsa, OK: Kane Miller, a division of EDC Publishing, 2018.

Internet Sites

Build Your Own Motorcycle
themotoexpert.com/how-to-build-a-motorcycle/

How to Build a Chopper Motorcycle
wikihow.com/Build-a-Chopper-Motorcycle

Motorcycle
kids.britannica.com/kids/article/motorcycle/400135

Index

bobbers, 4
brakes, 16

chrome, 18

engines, 6, 9, 14
exhaust systems, 9

fenders, 4, 12
foot pegs, 16
forks, 12
frames, 20

grips, 24

helmets, 6, 22

laws, 6, 22
lights, 26

rims, 20

seats, 28
shift controls, 16
sissy bars, 12
sprockets, 6

tires, 10, 12

About the Author

Lisa J. Amstutz is the author of more than 150 books for children. She enjoys reading and writing about science and technology. Lisa lives on a small farm in Ohio with her family.